Bioinformatics
Converting Data to Knowledge

A Workshop Summary by

Robert Pool, Ph.D. and Joan Esnayra, Ph.D.

Board on Biology
Commission on Life Sciences
National Research Council

NATIONAL ACADEMY PRESS
Washington, D.C.

NATIONAL ACADEMY PRESS · 2101 Constitution Avenue · Washington, D.C. 20418

NOTICE: The project that is the subject of this report was approved by the Governing Board of the National Research Council, whose members are drawn from the councils of the National Academy of Sciences, the National Academy of Engineering, and the Institute of Medicine. The members of the committee responsible for the report were chosen for their special competences and with regard for appropriate balance.

This report has been prepared with funds provided by the Department of Energy, grant DEFG02-94ER61939, and the National Cancer Institute, contract N01-OD-4-2139.

ISBN 0-309-07256-5

Additional copies are available from the National Academy Press, 2101 Constitution Ave., NW, Box 285, Washington, DC 20055; 800-624-6242 or 202-334-3313 in the Washington metropolitan area; Internet <http://www.nap.edu>.

Printed in the United States of America.

THE NATIONAL ACADEMIES

National Academy of Sciences
National Academy of Engineering
Institute of Medicine
National Research Council

The **National Academy of Sciences** is a private, nonprofit, self-perpetuating society of distinguished scholars engaged in scientific and engineering research, dedicated to the furtherance of science and technology and to their use for the general welfare. Upon the authority of the charter granted to it by the Congress in 1863, the Academy has a mandate that requires it to advise the federal government on scientific and technical matters. Dr. Bruce M. Alberts is president of the National Academy of Sciences.

The **National Academy of Engineering** was established in 1964, under the charter of the National Academy of Sciences, as a parallel organization of outstanding engineers. It is autonomous in its administration and in the selection of its members, sharing with the National Academy of Sciences the responsibility for advising the federal government. The National Academy of Engineering also sponsors engineering programs aimed at meeting national needs, encourages education and research, and recognizes the superior achievements of engineers. Dr. William A. Wulf is president of the National Academy of Engineering.

The **Institute of Medicine** was established in 1970 by the National Academy of Sciences to secure the services of eminent members of appropriate professions in the examination of policy matters pertaining to the health of the public. The Institute acts under the responsibility given to the National Academy of Sciences by its congressional charter to be an adviser to the federal government and, upon its own initiative, to identify issues of medical care, research, and education. Dr. Kenneth I. Shine is president of the Institute of Medicine.

The **National Research Council** was organized by the National Academy of Sciences in 1916 to associate the broad community of science and technology with the Academy's purposes of furthering knowledge and advising the federal government. Functioning in accordance with general policies determined by the Academy, the Council has become the principal operating agency of both the National Academy of Sciences and the National Academy of Engineering in providing services to the government, the public, and the scientific and engineering communities. The Council is administered jointly by both Academies and the Institute of Medicine. Dr. Bruce M. Alberts and Dr. William A. Wulf are chairman and vice chairman, respectively, of the National Research Council.

PLANNING GROUP FOR THE WORKSHOP ON BIOINFORMATICS: CONVERTING DATA TO KNOWLEDGE

DAVID EISENBERG, University of California, Los Angeles, California
DAVID J. GALAS, Keck Graduate Institute of Applied Life Sciences, Claremont, California
RAYMOND L. WHITE, University of Utah, Salt Lake City, Utah

Science Writer

ROBERT POOL, Tallahassee, Florida

Staff

JOAN ESNAYRA, Study Director
JENNIFER KUZMA, Program Officer
NORMAN GROSSBLATT, Editor
DEREK SWEATT, Project Assistant

Acknowledgments

The steering committee acknowledges the valuable contributions to this workshop of Susan Davidson, University of Pennsylvania; Richard Karp, University of California, Berkeley; and Perry Miller, Yale University. In addition, the steering committee thanks Marjory Blumenthal and Jon Eisenberg, of the NRC Computer Science and Telecommunications Board, for helpful input.

Preface

In 1993 the National Research Council's Board on Biology established a series of forums on biotechnology. The purpose of the discussions is to foster open communication among scientists, administrators, policy-makers, and others engaged in biotechnology research, development, and commercialization. The neutral setting offered by the National Research Council is intended to promote mutual understanding among government, industry, and academe and to help develop imaginative approaches to problem-solving. The objective, however, is to illuminate issues, not to resolve them. Unlike study committees of the National Research Council, forums cannot provide advice or recommendations to any government agency or other organization. Similarly, summaries of forums do not reach conclusions or present recommendations, but instead reflect the variety of opinions expressed by the participants. The comments in this report reflect the views of the forum's participants as indicated in the text.

For the first forum, held on November 5, 1996, the Board on Biology collaborated with the Board on Agriculture to focus on intellectual property rights issues surrounding plant biotechnology. The second forum, held on April 26, 1997, and also conducted in collaboration with the Board on Agriculture, was focused on issues in and obstacles to a broad genome project with numerous plant and animal species as its subjects. The third forum, held on November 1, 1997, focused on privacy issues and the desire to protect people from unwanted intrusion into their medical records. Proposed laws contain broad language that could affect biomedical and clinical research, in addition to the use of genetic testing in research.

After discussions with the National Cancer Institute and the Department of Energy, the Board on Biology agreed to run a workshop under the auspices of its forum on biotechnology titled "Bioinformatics: Converting Data to Knowledge" on February 16, 2000. A workshop planning group was assembled, whose role was limited to identifying agenda topics, appropriate speakers, and other participants for the workshop. Topics covered were: database integrity, curation, interoperability, and novel analytic approaches. At the workshop, scientists from industry, academe, and federal agencies shared their experiences in the creation, curation, and maintenance of biologic databases. Participation by representatives of the National Institutes of Health, National Science Foundation, US Department of Energy, US Department of Agriculture, and the Environmental Protection Agency suggests that this issue is important to many federal bodies. This document is a summary of the workshop and represents a factual recounting of what occurred at the event. The authors of this summary are Robert Pool and Joan Esnayra, neither of whom were members of the planning group.

This workshop summary has been reviewed in draft form for accuracy by individuals who attended the workshop and others chosen for their diverse perspectives and technical expertise in accordance with procedures approved by the NRC's Report Review Committee. The purpose of this independent review is to assist the NRC in making the published document as sound as possible and to ensure that it meets institutional standards. We wish to thank the following individuals, who are neither officials nor employees of the NRC, for their participation in the review of this workshop summary:

Warren Gish, Washington University School of Medicine
Anita Grazer, Fairfax County Economic Development Authority
Jochen Kumm, University of Washington Genome Center
Chris Stoeckert, Center for Bioinformatics, University of Pennsylvania

While the individuals listed above have provided many constructive comments and suggestions, it must be emphasized that responsibility for the final content of this document rests entirely with the authors and the NRC.

Joan Esnayra
Study Director

Contents

Dedication

This report is dedicated to the memory of Dr. G. Christian Overton for his vision and pioneering contributions to genomic research.

The Challenge of Information

Some 265 years ago, the Swedish taxonomist Carolus Linnaeus created a system that revolutionized the study of plants and animals and laid the foundation for much of the work in biology that has been done since. Before Linnaeus weighed in, the living world had seemed a hodge-podge of organisms. Some were clearly related, but it was difficult to see any larger pattern in their separate existences, and many of the details that biologists of the time were accumulating seemed little more than isolated bits of information, unconnected with anything else.

Linnaeus's contribution was a way to organize that information. In his *Systema Naturae,* first published in 1735, he grouped similar species—all the different types of maple trees, for instance—into a higher category called a genus and lumped similar genera into orders, similar orders into classes, and similar classes into kingdoms. His classification system was rapidly adopted by scientists worldwide and, although it has been modified to reflect changing understandings and interpretations, it remains the basis for classifying all living creatures.

The Linnaean taxonomy transformed biologic science. It provided biologists with a common language for identifying plants and animals. Previously, a species might be designated by a variety of Latin names, and one could not always be sure whether two scientists were describing the same organism or different ones. More important, by arranging biologic knowledge into an orderly system, Linnaeus made it possible for scientists to see patterns, generate hypotheses, and ultimately generate knowledge in a fundamentally novel way. When Charles Darwin published his *On the Origin of Species* in 1859, a century of Linnaean taxonomy had laid the groundwork that made it possible.

Today, modern biology faces a situation with many parallels to the one that Linnaeus confronted 2 1/2 centuries ago: biologists are faced with a flood of data that poses as many challenges as it does opportunities, and progress in the biologic sciences will depend in large part on how well that deluge is handled. This time, however, the major issue will not be developing a new taxonomy, although improved ways to organize data would certainly help. Rather, the major issue is that biologists are now accumulating far more data than they have ever had to handle before. That is particularly true in molecular biology, where researchers have been identifying genes, proteins, and related objects at an accelerating pace and the completion of the human genome will only speed things up even more. But a number of other fields of biology are experiencing their own data explosions. In neuroscience, for instance, an abundance of novel imaging techniques has given researchers a tremendous amount of new information about brain structure and function.

Normally, one might not expect that having too many data would be considered a problem. After all, data provide the foundation on which scientific knowledge is constructed, and the usual concern voiced by scientists is that they have too few data, not too many. But if data are to be useful, they must be in a form that researchers can work with and make sense of, and this can become harder to do as the amount grows.

Data should be easily accessible, for instance; if there are too many, it can be difficult to maintain access to them. Data should be organized in such a way that a scientist working on a particular problem can pluck the data of interest from a larger body of information, much of it not relevant to the task at hand; the more data there are, the harder it is to organize them. Data should be arranged so that the relationships among them are simple to understand and so that one can readily see how individual details fit into a larger picture; this becomes more demanding as the amount and variety of data grow. Data should be framed in a common language so that there is a minimum of confusion among scientists who deal with them; as information burgeons in a number of fields at once, it is difficult to keep the language consistent among them. Consistency is a particularly difficult problem when a data set is being analyzed, annotated, or curated at multiple sites or institutions, let alone by a well-trained individual working at different times. Even when analyses are automated to produce objective, consistent results, different versions of the software may yield differences in the results. Queries on a data set may then yield different answers on different days, even when superficially based on the same primary data. In short, how well data are turned into knowledge depends on how they are gathered, organized, managed, and exhibited—and those tasks are increasingly arduous as the data increase.

The form of the data that modern biologists must deal with is dramatically different from what Linnaeus knew. Then—and, indeed, at any point up until the last few decades—most scientific information was kept in "hard" format: written records, articles in scientific journals, books, artifacts, and various sorts of images, eventually including photographs, x-ray pictures, and CT scans. The information content changed with new discoveries and interpretations, but the form of the information was stable and well understood. Today, in biology and a number of other fields, the form is changing. Instead of the traditional ink on paper, an increasingly large percentage of scientific information is generated, stored, and distributed electronically, including data from experiments, analyses and manipulations of the data, a variety of images both real and computer-generated, and even the articles in which researchers describe their findings.

AN EXPLOSION OF DATABASES

Much of this electronic information is warehoused in large, specialized databases maintained by individuals, companies, academic departments in universities, and federal agencies. Some of the databases are available via the Internet to any scientist who wishes to use them; others are proprietary or simply not accessible online. Over the last decade, these databases have grown spectacularly in number, in variety, and in size. A recent database directory listed 500 databases just in molecular biology—and that included only publicly available databases. Many companies maintain proprietary databases for the use of their own researchers.

Most of the databases are specialized: they contain only one type of data. Some are literature databases that make the contents of scientific journals available over the Internet. Others are genome databases, which register the genes of particular species—human, mouse, fruit fly, and so on—as they are discovered, with a variety of information about the genes. Still others contain images of the brain and other body parts, details about the working of various cells, information on specific diseases, and many other subsets of biologic and medical knowledge.

Databases have grown in popularity so quickly in part because they are so much more efficient than the traditional means of recording and propagating scientific information. A biologist can gather more information in 30 minutes of sitting at a computer and logging in to databases than in a day or two of visiting libraries and talking to colleagues. But the more important reason for their popularity is that they provide data in a form that scientists can work with. The information in a scientific paper is intended only for viewing, but the data in a database have the potential to be downloaded, manipulated, analyzed, annotated, and combined with data from other databases. In short, databases can be far more than repositories—they can serve as tools for creating new knowledge.

A WORKSHOP IN BIOINFORMATICS

For that reason, databases hold the key to how well biologists deal with the flood of information in which they now find themselves awash. Getting control of the data and putting them to work will start with getting control of the databases. With that in mind, on February 16, 2000, the National Research Council's Board on Biology held a workshop titled "Bioinformatics: Converting Data to Knowledge." Bioinformatics is the emerging field that deals with the application of computers to the collection, organization, analysis, manipulation, presentation, and sharing of biologic data. A central component of bioinformatics is the study of the best ways to design and operate biologic databases. This is in contrast with the field of computational biology, where specific research questions are the primary focus.

At the workshop, 15 experts spoke on various aspects of bioinformatics, identifying some of the most important issues raised by the current flood of biologic data. The pages that follow summarize and synthesize the workshop's proceedings, both the presentations of the speakers and the discussions that followed them. Like the workshop itself, this report is not intended to offer answers as much as to pose questions and to point to subjects that deserve more attention.

The stakes are high—and not only for biologic researchers. "Our knowledge is not just of philosophic interest," said Gio Wiederhold, of the Computer Science department at Stanford University. "A major motivation is that we are able to use this knowledge to help humanity lead healthy lives." If the data now being accumulated are put to good use, the likely rewards will include improved diagnostic techniques, better treatments, and novel drugs—all generated faster and more economically than would otherwise be possible.

The challenges are correspondingly formidable. Biologists and their bioinformatics colleagues are in terra incognita. On the computer science side, handling the tremendous amount of data and putting them in a form that is useful to researchers will demand new tools and new strategies. On the biology side, making the most of the data will demand new techniques and new ways of thinking. And there is not a lot of time to get it right. In the time it takes to read this sentence, another discovery will have been made and another few million bytes of information will have been poured into biologic databases somewhere, adding to the challenge of converting all those data into knowledge.

Creating Databases

For most of the last century, the main problem facing biologists was gathering the information that would allow them to understand living things. Organisms gave up their secrets only grudgingly, and there were never enough data, never enough facts or details or clues to answer the questions being asked. Today, biologic researchers face an entirely different sort of problem: how to handle an unaccustomed embarrassment of riches.

"We have spent the last 100 years as hunter-gatherers, pulling in a little data here and there from the forests and the trees," William Gelbart, professor of molecular and cellular biology at Harvard University, told the workshop audience. "Now we are at the point where agronomy is starting and we are harvesting crops that we sowed in an organized fashion. And we don't know very well how to do it." "In other words," Gelbart said, "with our new ways of harvesting data, we don't have to worry so much about how to capture the data. Instead we have to figure out what to do with them and how to learn something from them. This is a real challenge."

It is difficult to convey to someone not in the field just how many data—and how many different kinds of data—biologists are reaping from the wealth of available technologies. Consider, for instance, the nervous system. As Stephen Koslow, director of the Office on Neuroinformatics at the National Institute of Mental Health, recounted, researchers who study the brain and nervous system are accumulating data at a prodigious rate, all of which need to be stored, catalogued, and integrated if they are to be of general use.

Some of the data come from the imaging techniques that help neuroscientists peer into the brain and observe its structure and function. Magnetic resonance imaging (MRI), computed tomography (CT), positron emission tomography (PET), and single-photon emission computed tomography (SPECT) each offer a unique way of seeing the brain and its components. Functional magnetic resonance imaging (fMRI) reveals which parts of a brain are working hardest during a mental activity, electroencephalography (EEG) tracks electric activity on the surface of the brain, and magnetoencephalography (MEG) traces deep electric activity. Cryosectioning creates two-dimensional images from a brain that has been frozen and carved into thin slices, and histology produces magnified images of a brain's microscopic structure. All of those different sorts of images are useful to scientists studying the brain and should be available in databases, Koslow said.

Furthermore, many of the images are most useful not as single shots but as series taken over some period. "The image data are dynamic data," Koslow said. "They change from day to day, from moment to moment. Many events occur in a millisecond, others in minutes, hours, days, weeks, or longer."

Besides images, neuroscientists need detailed information about the function of the brain. Each individual section of the brain, from the cerebral cortex to the hippocampus, has its own body of knowledge that researchers have accumulated over decades, Koslow noted. "And if you go into each of these specific regions, you will find even more specialization and detail—cells or groupings of cells that have specific functions. We have to understand each of these cell types and how they function and how they interact with other nerve cells."

"In addition to knowing how these cells interact with each other at a local level, we need to know the composition of the cells. Technology that has recently become available allows us to study individual cells or individual clusters of similar cells to look at either the genes that are being expressed in the cells or the gene products. If you do this in any one cell, you can easily come up with thousands of data points." A single brain cell, Koslow noted, may contain as many as 10,000 different proteins, and the concentration of each is a potentially valuable bit of information.

The brain's 100 billion cells include many types, each of which constitutes a separate area of study; and the cells are hooked together in a network of a million billion connections. "We don't really understand the mechanisms that regulate these cells or their total connectivity," Koslow said; "this is what we are collecting data on at this moment."

Neuroscientists describe their findings about the brain in thousands of scientific papers each year, which are published in hundreds of journals. "There are global journals that cover broad areas of neuroscience research," Koslow said, "but there are also reductionist journals that go from specific areas—the cerebral cortex, the hippocampus—down to the neuron, the synapse, and the receptor."

The result is a staggering amount of information. A single well-studied substance, the neurotransmitter serotonin, has been the subject of 60,000-70,000 papers since its discovery in 1948, Koslow said. "That is a lot of information to digest and try to synthesize and apply." And it represents the current knowledge base on just one substance in the brain. There are hundreds of others, each of which is a candidate for the same sort of treatment.

FOUR ELEMENTS OF A DATABASE

"We put four kinds of things into our databases," Gelbart said. "One is the biologic objects themselves"—such things as genetic sequences, proteins, cells, complete organisms, and whole populations. "Another is the relationships among those objects," such as the physical relationship between genes on a chromosome or the metabolic pathways that various proteins have in common. "Then we also want classifiers to help us relate those objects to one another." Every database needs a well-defined vocabulary that describes the objects in it in an unambiguous way, particularly because much of the work with databases is done by computers. Finally, a database generally contains metadata, or data about the data: descriptions of how, when, and by whom information was generated, where to go for more details, and so on. "To point users to places they can go for more information and to be able to resolve conflicts," Gelbart explained, "we need to know where a piece of information came from."

Creating such databases demands a tremendous amount of time and expertise, said Jim Garrels, president and CEO of Proteome, Inc., in Beverly, Massachusetts. Proteome has developed the Bioknowledge Library, a database that is designed to serve as a central clearinghouse for what researchers have learned about protein function. The database contains descriptions of protein function as reported in the scientific literature, information on gene sequences and protein structures, details about proteins' roles in the cell and their interactions with other proteins, and data on where and when various proteins are produced in the body.

DATABASE CURATION

It is a major challenge, Garrels said, simply to capture all that information and structure it in a way that makes it useful and easily accessible to researchers. Proteome uses a group of highly trained curators who read the scientific literature and enter important information into the database. Traditionally, many databases, such as those on DNA sequences, have relied on the researchers themselves to enter their results, but Garrels does not believe that would work well for a database like Proteome's. Much of the value of the database lies in its curation—in the descriptions and summaries of the research that are added to the basic experimental results. "Should authors curate their own papers and send us our annotation lines? I don't think so. We train our curators a lot, and to have 6,000 untrained curators all sending us data on yeast would not work." Researchers, Garrels said, should deposit some of their results directly into databases—genetic sequences should go into sequence databases, for instance—but most of the work of curation should be left to specialists.

In addition to acquiring and arranging the data, curators must perform other tasks to create a workable database, said Michael Cherry, technical manager for Stanford University's Department of Genetics and one of the specialists who developed the Saccharomyces Genome Database and the Stanford Microarray Database. For example, curators must see that the data are standardized, but not too standardized. If computers are to be able to search a database and pick out the information relevant to a researcher's query, the information must be stored in a common format. But, Cherry said, standardization will sometimes "limit the fine detail of information that can be stored within the database."

Curators must also be involved in the design of databases, each of which is customized to its purpose and to the type of data; they are responsible for making a database accessible to the researchers who will be using it. "Genome databases are resources for tools, as well as resources for information," Cherry said, in that the databases must include software tools that allow researchers to explore the data that are present.

In addition, he said, curators must work to develop connections between databases. "This is not just in the sense of hyperlinks and such things. It is also connections with collaborators, sharing of data, and sharing of software."

Perhaps the most important and difficult challenge of curation is integrating the various sorts of data in a database so that they are not simply separate blocks of knowledge but instead are all parts of a whole that researchers can work with easily and efficiently without worrying about where the data came from or in what form they were originally generated.

"What we want to be able to do," Gelbart said, "is to take the structural information that is encapsulated in the genome—all the gene products that an organism encodes, and the instruction manual on how those gene products are deployed—and then turn that into useful information

The Need for Bioinformaticists

As the number and sophistication of databases grow rapidly, so does the need for competent people to run them. Unfortunately, supply does not seem to be keeping up with demand.

"We have a people problem in this field," said Stanford's Gio Wiederhold. "The demand for people in bioinformatics is high at all levels, but there is a critical lack of training opportunities and also of available trainees."

Wiederhold described several reasons for the shortage of bioinformatics specialists. People with a high level of computer skills are generally scarce, and "we are competing with the excitement that is generated by the Internet, by the World Wide Web, by electronic commerce." Furthermore, biology departments in universities have traditionally paid their faculty less than computer-science or engineering departments. "That makes it harder for biologists and biology departments to attract the right kind of people."

Complicating matters is the fact that bioinformatics specialists must be competent in a variety of disciplines—computer science, biology, mathematics, and statistics. As a result, students who want to enter the field often have to major in more than one subject. "We have to consider the load for students," Wiederhold said. "We can't expect every student interested in bioinformatics to satisfy all the requirements of a computer-science degree and a biology degree. We have to find new programs that provide adequate training without making the load too high for the participants."

Furthermore, even those with the background and knowledge to go into bioinformatics worry that they will find it difficult to advance in such a nontraditional specialty. "The field of bioinformatics is scary for many people," Wiederhold said. "Because it is a multidisciplinary field, people are worried about where the positions are and how easily they will get tenure." Until universities accept bioinformatics as a valuable discipline and encourage its practitioners in the same way as those in more traditional fields, the shortage of qualified people in the field will likely continue.

that tells us about the biologic process and about human disease. On one pathway, we are interested in how those gene products work—how they interact with one another, how they are expressed geographically, temporally, and so on. Along another path, we would like to study how, by perturbing the normal parts list or instruction manual, we create aberrations in how organisms look, behave, carry out metabolic pathways, and so on. We need databases that support these operations."

One stumbling block to such integration, Gelbart said, is that the best way to organize diverse biologic data would be to reflect their connections in the body. But, he said, "we really don't understand the design principles, so we don't know the right way to do it." It is a chicken-and-egg problem of the sort that faced Linnaeus: A better understanding of the natural world can be expected to flow from a well-organized collection of data, but organizing the data well demands a good understanding of that world. The solution is, as it was with Linnaeus, a bootstrap approach: Organize the data as well as you can, use them to gain more insights, use the new insights to further improve the organization, and so on.

Barriers to the Use of Databases

If researchers are to turn the data accumulating in biologic databases into useful knowledge, they must first be able to access the data and work with them, but this is not always as easy as it might seem. The form in which data have been entered into a database is critical, as is the structure of the database itself, yet there are few standards for how databases should be constructed. Most databases have sprung up willy-nilly in response to the special needs of particular groups of scientists, often with little regard to broader issues of access and compatibility. This situation seriously limits the usefulness of the biologic information that is being poured into databases at such a prodigious rate.

PROPRIETARY ISSUES

The most basic barrier to putting databases to use is that many of them are unavailable to most researchers. Some are proprietary databases assembled by private companies; others are collections that belong to academic researchers or university departments and have never been put online. "The vast majority of databases are not actually accessible through the Internet right now," said Peter Karp, director of the Bioinformatics Research Group at SRI International in Menlo Park, California. If a database cannot be searched online, few researchers will take advantage of it even if, in theory, the information in it is publicly available. And even the hundreds of databases that can be accessed via the Internet are not necessarily easy to put to work. The barriers come in a number of forms.

One problem is simply finding relevant data in a sea of information, Karp said. "If there are 500 databases out there, at least, how do we know which ones to go to, to answer a question of interest?" Fortunately for biologists, some locator help is available, noted Douglas Brutlag, professor of biochemistry and medicine at Stanford University. A variety of database lists are available, such as the one published in the *Nucleic Acid Research* supplemental edition each January, and researchers will find the large national and international databases—such as NCBI, EBI, DDBJ, and SWISS-PROT—to be good places to start their search. "They often have pointers to where the databases are," Brutlag noted. Relevant data will more than likely come from a number of different databases, he added. "To do a complete search, you need to know probably several databases. Just handling one isn't sufficient to answer a biologic question." The reason lies in the growing integration of biology, Karp said. "Many databases are organized around a single type of experimental data, be it nucleotide-sequence data or protein-structure data, yet many questions of interest can be answered only by integrating across multiple databases, by combining information from many sources."

The potential of such integration is perhaps the most intriguing thing about the growth of biologic databases. Integration holds the promise of fundamentally transforming how biologic research is done, allowing researchers to synthesize information and make connections among many types of experiments in ways that have never before been possible; but it also poses the most difficult challenge to those who develop and use the databases. "The problem," Karp explained, "is that interaction with a collection of databases should be as seamless as interaction with any single member of the collection. We would like users to be able to browse a whole collection of databases or to submit complex queries and analytic computations to a whole collection of databases as easily as they can now for a single database." But integrating databases in this way has proved exceptionally difficult because the databases are so different.

"We have many disciplines, many subfields," said Gio Wiederhold, of Stanford University's Computer Science Department, "and they are autonomous—and must remain autonomous—to set their own standards of quality and make progress in their own areas. We can't do without that heterogeneity." At the same time, however, "the heterogeneity that we find in all the sources inhibits integration." The result is what computer scientists call "the interoperability problem," which is actually not a single difficulty, but rather a group of related problems that arise when researchers attempt to work with multiple databases. More generally, the problem arises when different kinds of software are to be used in an integrated manner.

DISPARATE TERMINOLOGY

The simplest yet most unyielding difficulty is that biologists in different specialties tend to speak somewhat different languages. They use jargon and terminology peculiar to their own subfields, and they have their own particular theories and models underlying the collection of data. "We get major terminologic problems," Wiederhold said, "because the terms used in one field will have different granularity depending on the level at which the abstractions or concepts in that field work and will have different scope, so a term taken in a different context often has a somewhat different meaning. The simple solution is that we will make everybody speak the same language. That, however, requires a degree of stability that we cannot expect in any technology and certainly not in bioinformatics. The fields are moving rapidly—new terms will develop, meanings of terms will change—so we will have to deal with the difference in terminology and recognize that there are differences and be careful with precision."

INTEROPERABILITY

Besides the differing terminologies, someone who wishes to work across many databases must also deal with differences in how the various collections structure their data. "There are many protein databases out there," Karp said, "and each one chooses to conceptualize or represent proteins in its schema in a different way. So someone who wants to issue a query to 10 protein databases has to examine each database to figure out how it encodes a protein, what information it encodes, what field names it uses, and what units of measurement it uses. There are also different data models: object-data models versus relational-data models versus ad hoc, invented-by-the-database-author data models. Daniel Gardner, of Cornell University, added, "it is interfaces, not uniformity, that can provide interoperability—interfaces for data exchange and data-format description, interfaces to recognize data-model intersections, to exchange metadata and to parse queries."

Wiederhold continued, "Another very important issue is the heterogeneity in user expertise. Addressing complex queries to large collections of databases requires significant sophistication in the user who is going to create a query of that form. The vast majority of users simply do not have that expertise today."

None of those issues is new, and for a number of years bioinformatics specialists have been devising ways to improve interoperability. Beginning in 1994, Karp organized a series of workshops on interconnecting molecular-biology databases. Those workshops stimulated the development of a number of practical software tools. "I am pleased to report," Karp said, "that over the last 5 years there really has been some significant progress in building a software infrastructure for database interoperation, which we can liken to building the Internet. Just as the Internet connects a diverse set of geographically distributed locations, we have seen growth in a software infrastructure for connecting molecular-biology databases."

Bioinformatics specialists have developed two broad approaches to integrating databases, each with its strengths and weaknesses. The first, which Karp referred to as the warehousing approach, combines a large number of individual databases in a single computer and lets outside users submit queries to that collection of databases. An example is the Sequence Retrieval System (SRS), which contains 133 databases and is available through the European Bioinformatics Institute (EBI). The SRS treats all the files in all the databases as text files and indexes the databases by keywords within the files and by record names in each of the fields within each database. People using the system search for relevant files by keyword and by record name. "The main advantage of the text warehousing approach," Karp said, "is that users can essentially use point-and-click. You enter a set of keywords and you get back lots and lots of records that match those keywords. Point-and-click is the major advantage of this approach because it is easy for people to use, but it is also the major disadvantage because it can take so long to evaluate complex queries."

Suppose, for example, that someone wished to find examples of sets of genes that were clustered tightly on a single chromosome and that specified enzymes that worked within a single metabolic pathway. The search would demand the comparison of two types of information: on the location of genes and on the metabolic pathways that particular enzymes play a role in. To perform that search in the SRS, Karp said, "we might enter a keyword like *pathway* and get back the names of every pathway and every pathway database within the SRS. To answer this query and to find linked genes in a single metabolic pathway, we would have to point-and-click through hundreds of pathway records, follow each pathway to its enzymes, and follow each enzyme to its genes. We would have a case of repetitive-stress injury by the time we were finished."

The second system for integrating databases is the multidatabase approach, which takes a query from a user, distributes the query via the Internet to a set of many databases, and then collects and displays the results. Examples of that approach are the Kleisli/K2 system developed by Chris Overton and colleagues at the University of Pennsylvania, the OPM system developed by Victor Markowitz at Gene Logic, and the TAMBIS system (which is built on Kleisli) developed by Andy Brass and

Carole Goble at the University of Manchester. Because the individual databases maintain their structures instead of being treated as collections of text files, searches can be much more powerful and exact in this type of system than in the warehousing system. After the user formulates a question, a query processor transforms the question into individual queries sent to whichever of the various member databases might have information relevant to the original question. Later, the query engine receives and integrates the results of the individual queries and returns the results to the user. "For instance," Karp said, "in our pathway-and-gene example, the query processor might farm out individual queries across the Internet, combine the results, formulate more queries for genome databases, and then combine the results. The main advantage of the multidatabase and warehousing approaches is that they are high-throughput approaches. They allow us to process complex queries that might access tens of databases and thousands or tens of thousands of objects to perform interesting system-level analyses of large amounts of data. [For text-based warehousing], the point-and-click approach will never do that."

In contrast, although anyone can point-and-click with little training, the preparation of the complex queries for a multidatabase system demands much greater expertise. "The majority of the multidatabase systems force their users to learn some complex query language," Karp said. "They also force their users to learn a lot about the schemes of each database they want to query. Some graphical query interfaces are available, but they tend to be fairly primitive. More work is needed in this direction."

In short, the good news is that systems do exist to allow researchers to search 100 or more databases simultaneously. The bad news is that it is still difficult for anyone but database experts to perform the sorts of complex searches that are most valuable to researchers. And further bad news, Karp said, is that many of the existing databases cannot be integrated into such interoperation systems, because they do not have the necessary structure. "Many individual databases have not been constructed with any kind of database-management system. They are simply text files created with a text editor. Many have no defined ontology or schema, so it is difficult to tell what data are in them and what the different fields mean. Most are not organized according to any standard data model, and many of these flat files have an irregular structure that is very hard to parse. They often have inconsistent semantics."

The take-home message, Karp said, is that databases should be constructed with an eye to interoperability, but, so far, most are not. "Unfortunately, database expertise is very much lacking in the vast majority of bioinformatics database projects. In general, these projects have been lacking in the discipline to use database-management systems, to use more standardized data models, and to come up with a more-regular syntax." The result is that only a minority of all biology databases are available over the Internet for the interoperation engines to use. In the future, advances in tools designed for the World Wide Web, in combination with advances in databases and in other forms of software, are likely to make more biology data more easily available. This will involve progress in multiple components of computer science and attention to the specific interests of biologists as data generators and users. It will also require biologists to present their needs in ways that excite database experts and other computer scientists to overcome the expertise scarcity noted by Karp.

Maintaining the Integrity of Databases

Databases can contain billions of bytes of information, so it is inevitable that some errors will creep in. Nonetheless, the researchers who work with databases would like to keep the errors to a minimum. Error prevention and correction must be integral parts of building and maintaining databases.

The reasons for wanting to minimize errors are straightforward, said Chris Overton, director of the Center for Bioinformatics at the University of Pennsylvania. "I work on genome annotation, which, broadly speaking, is the analysis and management of genomic data to predict and archive various kinds of biologic features, particularly genes, biologic signals, sequence characteristics, and gene products." Presented with a gene of unknown function, a gene annotator will look for other genes with similar sequences to try to predict what the new gene does. "What we would like to end up with," Overton explained, "is a report about a genomic sequence that has various kinds of data attached to it, such as experimental data, gene predictions, and similarity to sequences from various databases. To do something like this, you have to have some trusted data source." Otherwise, the researchers who rely on the genome annotation could pursue false trails or come to false conclusions.

Other fields have equally strong reasons for wanting the data in databases to be as accurate as possible. It is generally impractical or impossible for researchers using the data to check their accuracy themselves: if the data on which they base their studies are wrong, results of the studies will most likely be wrong, too.

ERROR PREVENTION

To prevent errors, Overton commented, it is necessary first to know how and why they appear. Some errors are entry errors. The experimentalists who generate the data and enter them into a database can make mistakes in their entries, or curators who transfer data into the database from journal articles and other sources can reproduce them incorrectly. It is also possible for the original sources to contain errors that are incorporated into the database. Other errors are analysis errors. Much of what databases include is not original data from experiments, but information that is derived in some way from the original data, such as predictions of a protein's function on the basis of its structure. "The thing that is really going to get us," Overton said, "is genome annotation, which is built on predictions. We have already seen people taking predictions and running with them in ways that perhaps they shouldn't. They start out with some piece of genomic sequence, and from that they predict a gene with some sort of ab initio gene-prediction program. Then they predict the protein that should be produced by that gene, and then they want to go on and predict the function of that predicted protein." Errors can be introduced at any of those steps.

Once an error has made it into a database, it can easily be propagated—not only around the original database, but into any number of other systems. "Computational analysis will propagate errors, as will transformation and integration of data from various public data resources," Overton said. "People are starting to worry about this problem. Data can be introduced in one database and then just spread out, like some kind of virus, to all the other databases out there." And as databases become more closely integrated, the problem will only get worse.

ERROR CORRECTION

Because Overton's group is involved with database integration, taking information from a number of databases and combining it in useful ways, it has been forced to find ways to detect and fix as many errors as possible in the databases that it accesses. For example, it has developed a method for correcting errors in the data that it retrieves from GenBank, the central repository for gene sequences produced by researchers in the United States and around the world. "Using a rule base that included a set of syntactic rules written as grammar," Overton said, "we went through all the GenBank entries for eukaryotic genes and came up with a compact representation of the syntactic rules that describe eukaryotic genes." If a GenBank entry was not "grammatical" according to this set of syntactic rules, the system would recognize that there must be an error and often could fix it.

"That part was easy", he said. "The part that got hard was when we had to come up with something like 20 pages of expert-system rules to describe all the variations having to deal with the semantic variability that was in GenBank. At the moment, the system—which we have been working on for a relatively long time—can recognize and correct errors in about 60–70% of GenBank entries. Another 30-40% we end up having to repair by hand. Unfortunately, although dealing with feature-table information in GenBank is relatively easy—the information is highly structured, and you can write down rules that capture the relationships between all the features—that is certainly not true for a lot of the other biologic data we are looking at, and we do not have any way to generalize these error-detection protocols for other kinds of data that are out there." In short, even in the best case, it is not easy to correct errors that appear in databases; and in many cases, there is no good way to cleanse the data of mistakes.

On the basis of his group's experience with detecting and correcting errors, Overton offered a number of lessons. The first and simplest is that it is best not to let the errors get into the database in the first place. "Quality control and quality assurance should be done at the point of entry—that is, when the data are first entered into a database. We shouldn't have had to run some tool like this. It should have been run at GenBank at the time the information was entered. That would have been a way to clear up a lot of the errors that go in the database." Supporting this comment was Michael Cherry of Stanford, who stated that "everything we do has to be right. The quality control has to be built into the design."

THE IMPORTANCE OF TRAINED CURATORS AND ANNOTATORS

A related piece of advice was that the best people to enter the data are not the researchers who have generated them, but rather trained curators. "At GenBank, data are entered by the biologists who determine a sequence. They are not trained annotators; but when they deposit the nucleic acid sequence in GenBank, they are required to add various other information beyond the sequence data. They enter metadata, and they enter features, which are the equivalent of annotations. That is why we get a lot of errors in the database. Most of the people involved in this process come to the same conclusion: that trained annotators give generally higher quality and uniformity of data than do scientists. So one goal would be to just get the biologist out of the loop of entering the data."

Once errors have crept into a database, Overton said, there is likely to be no easy way to remove them. "Many of the primary databases are not set up to accept feedback. When we find errors in, say, GenBank, there is nobody to tell." Without a system in place to correct mistakes, those who operate a database have a difficult time learning about errors and making corrections. "Furthermore," Overton said, "we would get no credit for it even if we did supply that kind of information." Scientists are rewarded for generating original data but not for cleaning up someone else's data. "The other part of the problem is that most of these databases do a very poor job of notifying you when something has changed. It is extremely difficult to go to GenBank and figure out whether something has changed." So even if an error is discovered and corrected in a database, anyone who has already downloaded the erroneous data is unlikely to find out about the mistake.

DATA PROVENANCE

One way to ameliorate many of the problems with errors in databases, Overton said, is to keep detailed records about the data: where they came from, when they were deposited, how one gets more information about them, and any other relevant details concerning their pedigree. The approach is called "data provenance," and it is particularly applicable to minimizing errors in data that propagate through various databases.

"The idea of data provenance," Overton said, "is that databases should describe the evidence for each piece of data, whether experimental or predicted, so you should know where the data came from. Furthermore, you should be able to track the source of information. In our own internal databases, we track not only the history of everything as it changes over time, but all the evidence for every piece of data. If they change, we have a historical record of it."

"This is an important and extremely difficult problem," Overton said. "There is no general solution for it at the moment."

DATABASE ONTOLOGY

At Knowledge Bus, Inc., in Hanover, Maryland, Bill Andersen has a different approach to dealing with records. Knowledge Bus is developing databases that incorporate ontologic theories—theories about the nature of and relationships among the various types of objects in a database. In other words, the databases "know" a good deal about the nature of the data that they contain and what to expect from them, so they can identify various errors simply by noting that the data do not perform as postulated.

For example, one of the thousands of axioms that make up the Knowledge Bus ontology describes what to expect from two closely related reactions, one of which is a specialized version of the other, such as glucose phosphorylation and glucose phosphorylation catalyzed by the enzyme hexokinase. Those two reactions are identical except that the second proceeds with the help of an enzyme. "The rule," Andersen explained, "just explains that if the normal glucose phosphorylation has a certain free energy, then the one catalyzed by hexokinase will have the same free energy." (Free energy, a concept from thermodynamics, is related to the amount of work performed, or able to be performed, by a system.)

Suppose that the system pulls in perhaps from one or more databases on the Internet experimentally determined values for the free energy of glucose phosphorylation and of glucose phosphorylation catalyzed by hexokinase, and suppose further that they do not agree. The system immediately recognizes that something is wrong and, equally important, has a good starting point for figuring out exactly what is wrong and why. "What went wrong had a lot to do with the theory you built the database with," Andersen said. "Either the constraints are wrong, the ontology is wrong, or the data are wrong. In any case, we can use the violated constraint to provide an explanation. Here is our starting point. This is what went wrong. The idea that I want to get across is that once we have got hold of that proof [that an error has occurred], we can start to look at the information. All the proof told us is that, according to our model, the database is wrong. But how? Was the information input reliable? Was another class of mistake made? What can we tell from examining the provenance of the information? Maybe we believe this piece more than that piece because of where they came from, and we can resolve it that way."

The ontology is combined with extensive metadata (data on the data) so curators can quickly learn where data came from and what their potential weaknesses are. "Combining the annotations and the proof," Andersen said, "we can start reasoning about the errors that have appeared. We can use these facilities to provide tools to guide human curators to the sources of error so that they can fix them rapidly." Once an error has been identified, a curator can use the information about it to decide what to do. "You can remove the conflicting data, or you can simply take the formula and put an annotation on it, say that it is in conflict and we don't know what to do with it. Mark it and go on. That is also possible to do. You don't have to eliminate the conflict from the database."

Either way, Andersen said, by having a database that can identify inconsistencies in data and give information about how and why they are inconsistent, curators should be able to deal with errors much more effectively than is possible with current database configurations.

Maintaining Privacy

As databases become increasingly widespread, more and more people will find that data about them appear in databases. The data might have been gathered as part of an experiment or might represent information collected by doctors during normal medical care of patients; they could include genetic information, medical histories, and other personal details. But whatever their form, warned Stanford's Gio Wiederhold, those who work with databases must be careful to respect the privacy and the concerns of the people whose data appear in them.

"You have to be very careful about how people will feel about your knowledge about them," he said. Detailed medical information is a sensitive subject, but genetic information may well be even touchier. Genetic data can be used for paternity testing, for detecting the presence of genetic diseases, and eventually for predicting a person's physical and psychologic propensities. "Privacy is very hard to formalize and doesn't quite follow the scientific paradigm that we are used to. That doesn't mean that it is not real to people—perceptions count here. I request that scientists be very sensitive to these kinds of perceptions, make every possible effort to recognize the problems that they entail, and avoid the backlash that can easily occur if privacy is violated and science is seen in a negative light."

There are also a number of practical issues in preserving privacy, Wiederhold noted, such as the possibility of unethical use of genetic information by insurance companies. Methods for protecting privacy have not kept pace with the increasing use of shared databases.

"In our work, we are always collaborating," Wiederhold said, "but the technical means that we have today for guarding information come from commerce or from the military and are quite inadequate for protecting collaboration." In those other fields, the first line of defense has been to control access and to keep all but a select few out of a database altogether. That won't work in research: "We have to give our collaborators access."

Those who run databases that contain sensitive information will therefore need to find different approaches to protecting privacy. "We have to log and monitor what gets taken out. It might also be necessary to ensure that some types of information go out only to those who are properly authorized," he said, noting the well-reported case of a person who logged onto an Internet music site and, instead of downloading a music track, downloaded the credit-card numbers of hundreds of thousands of the site's customers. "They obviously were not checking what people were taking out. The customer had legitimate access, but he took out what he shouldn't have taken out."

Wiederhold concluded: "Unless we start logging the information that is taken out, and perhaps also filtering, we will not be fulfilling our responsibilities."

Converting Data to Knowledge

Ultimately, the tremendous amount of information now being generated by biologists and deposited into databases is useful only if it can be applied to create knowledge. And, indeed, researchers are finding that the many databases now available are making it possible for them to do many things that they never could before.

DATA MINING

Perhaps the best-known technique is data mining. Because many data are now available in databases—including information on genetic sequences, protein structure and function, genetic mutations, and diseases—and because data are available not only on humans but also on many other species, scientists are finding it increasingly valuable to "mine" the databases for patterns or connected bits of information that can be assembled into a larger picture. By integrating details from various sources in this way, researchers can generate new knowledge from the data assembled in the databases.

Much of today's data mining is done by biologists who have discovered a new gene or protein and wish to figure out what it does, said Stanford's Douglas Brutlag, professor of biochemistry and medicine. At first, the researcher might know little more about the new find than its genetic sequence (for a gene) or its sequence of amino acids (for a protein), but often that is enough. By searching through databases to find similar genes or proteins whose functions have already been identified, the researcher might be able to determine the function of the new item or at least make a reasonable guess.

In the simplest cases, data mining might work like this: A genome scientist has a new, unidentified human gene in hand and proceeds to search through genome databases on other species—the mouse, the fruit fly, the worm *Caenorhabditis elegans*, and so on—looking for known genes with a similar genetic sequence. Different species share many of the same genes; although the sequence of a particular gene might vary from species to species (more for distantly related species than for closely related ones), it is generally feasible to pick out genes in different species that correspond to a particular gene of interest. If a database search turns up such correspondences, the researcher now has solid evidence about what the newly discovered gene might do.

In reality, the database analysis of genes and proteins has become far more sophisticated than that simple searching for "homologues," or items with similar structures. For instance, Brutlag noted, researchers have developed databases of families of sequences in which each family consists of a group of genes or proteins that have a structure or function in common. When a new gene or protein is found, its discoverer can compare it not just one on one with other individual genes or proteins, but with entire families, looking for one in which it fits. This is a more powerful technique than one-to-one comparisons because it relies on general patterns instead of specific details. Just as an unusual-looking mutt can be identified as a dog even if it cannot be classified as a particular breed, a new protein can often be placed in a family of proteins even if it is not a homologue of any known protein.

Researchers have developed a series of databases that can be used to classify genes and proteins, each with a different technique for identifying relationships: sequence motifs, consensus sequences, position-specific scoring matrices, hidden Markov models, and more. "I can hardly keep up with the databases myself," Brutlag said. With these techniques, researchers can now usually determine what a newly discovered human gene or protein does on the basis of nothing more than the information available in databases, Brutlag said. About a year before the workshop, his group created a database of all known human proteins and their functions. Over the next year, each time a new human protein was analyzed, they analyzed it by using homologues and a technique developed in Brutlag's laboratory called eMATRICES. "Using both methods, we assigned biologic functions to almost 77% of the human proteins. More than three-fourths of new proteins could be characterized by a technician who never left his computer; although the ultimate test remains experimental verification, this method promises to speed up drug discovery, for example."

INTERNATIONAL CONSORTIUM FOR BRAIN MAPPING

A different way of exploiting the information in biologic databases is demonstrated by the International Consortium for Brain Mapping. The consortium is developing a database that will provide physicians and neuroscientists with a description of the structure and function of the human brain that is far more accurate and complete than anything available today. The database will be a combination of brain atlas, visualization device, clinical aid, and research tool.

Mapping the human brain is complicated, and not simply because the brain is a complicated organ. The more important factor is that the brain varies from person to person. "Every brain is different in structure and probably more so in function," said John Mazziotta, director of the Brain Mapping Division at the UCLA School of Medicine. Even identical twins have brains that look noticeably different and whose functions are localized in slightly different areas. The brains of unrelated people have even greater variation, and this makes it impossible to create a single, well-defined representation of the brain. Any representation must be probabilistic—that is, the representation will not describe exactly where each structure or function lies, but will instead provide a set of possible locations and the likelihood of each. So instead of creating a single, sharply defined map laying out the various features of the human brain and coloring in the areas responsible for different functions, any brain-mapping project must find some way to capture and display the inherent fuzziness in where things lie in the brain.

"That is very hard," Mazziotta said. "In fact, we don't yet have a good way to do it." Nonetheless, the consortium has developed ways in which the natural variation from brain to brain can be captured and displayed, which make it possible to get a much clearer picture of what is normal in the human brain and what falls outside the normal range.

The desire to create a brain-mapping tool was motivated by two main factors, Mazziotta said. The first was the sense that the various researchers in the field of brain mapping were heading off in many directions and that no one was attempting to bring all the threads together to see what they were jointly producing. "As in a pointillist painting, all of us in the imaging field were working on our dot in isolation, trying to refine it and get it better. The concept was that if we worked together and pooled the data, we would have a composite image that would show the big picture and be much more than the sum of the individual points."

One experiment in particular was a major factor behind the push to create the brain-mapping consortium, Mazziotta said. He and colleagues at UCLA and in London studied the brains of four subjects who were observing an object moving across their visual field. The researchers found that in each subject a particular small area of the brain became active during the experiment and the researchers could identify the area as being involved in the visual perception of motion. "The location was consistent across subjects, but we don't know how consistent, because there is such variance [in brain structure] between individuals. This is a big problem." Without having a good idea of what constitutes normal variation in brain structure and function among individuals, the researchers had no way to judge the meaning of their results.

One of Mazziotta's collaborators, John Watson, combed through the literature in search of information on patients who lose their ability to detect motion in their visual fields. He eventually found an article describing such a patient; the patient had damage in exactly the part of the brain that the researchers had already zeroed in on. Watson also found a 1918 description of which parts of a newborn's brain are myelinated—that is, in which areas the neurons were sheathed with myelin, a fatty coating that improves the performance of nerve cells. Only a few primary parts of the brain are myelinated at birth, but Watson found that one of those sections correlated precisely with the part of the brain that seemed to detect motion in the visual field. "Newborn infants might want to know that something is coming at them really fast in their visual environment," Mazziotta said, "so that area has to be ready to go at birth. This is speculation, but it makes sense."

At the end of the process, the group of researchers had woven together evidence from a number of studies that this particular spot in the brain was responsible for detecting motion in the visual field. "The only problem was that this was a library exercise," Mazziotta concluded. "What it needs to be—and what we want it to be—is a digital database exercise, where the framework is the structure of human brain, so we can do an experiment, find this observation, and go deep into the data and find other features that are now very awkward to identify."

The second motivating factor for developing the brain-atlas database, Mazziotta said, was the sheer amount of data generated by even the simplest experiments with the human brain. "A typical human male brain has a volume of about 1,500 cubic centimeters, and any given cell can express, at any time, 50,000-75,000 gene products. If you took the most crude sampling—1 cubic centimeter—that one could envision, that represents 75 million data points. If you scale it down to a cellular size, 10 micrometers [10 thousandths of a millimeter]—that represents 75,000 trillion data points for one brain at one time. If you take that across the age range—from birth to 100 years—and do that for different populations, you get truly astronomical amounts of data, just for this one perspective on gene expression as a function of age and spatial resolution." With the potential for so many data, it seemed important to establish a place that could deal with them effectively, integrating the various types of data and creating a representation of the brain that was as complete as possible.

The brain-mapping consortium contains sites around the world, including the United States, Japan, and Scandinavia. Ultimately, it will include data on 7,000 subjects, although data on only 500 have been collected so far. The data include not only a variety of brain images both structural and functional, but also histories of the subjects, demographic information, behavioral information from handedness to neuropsychology, and, for most of the subjects, DNA samples. Mazziotta said that the system makes it possible to study the relationships among genetics, behavior, brain structure, and brain function in a way that takes into account the variations in structure and function that occur among people.

Mazziotta offered three examples of how this sort of system can be put to work. In the first, researchers at UCLA looked at images of the brains of 14-year-olds and 20-year-olds and asked whether there were any differences—a question that, because of natural brain-to-brain variation, would be nearly impossible to answer by looking at one or two subjects at each age. "The prevailing wisdom was that there was not a lot of change in brain structure between those ages." But by mapping the normal range of 14-year-old brains and the normal range of 20-year-old brains, the group showed that changes did indeed take place in the prefrontal cortex and the base of the forebrain.

A second study compared the brains of a population of patients who had early Alzheimer's disease, averaged in probabilistic space, with the brains of a population of patients in the later stages of the disease. It found that Alzheimer's disease causes changes in the gross structure of the brain, thinning the corpus callosum and causing the upper part of the parietal lobe to shrink. "This is an example of a disease demonstrated not in an individual but in a group," Mazziotta said, "and it is useful clinically to evaluate different therapies." One might, for example, perform a clinical trial in which one-third of the patients were given an experimental therapy, another third a conventional therapy, and the rest a placebo. At the end of the trial, the probabilistic brain-mapping technique would produce a measure of the changes that took place in the brains of the three groups of patients and offer an objective measure of how well the different therapies worked.

The final example was a diagnostic one. "Let's say that a 19-year-old woman has seizures that come from this part of the brain in the frontal lobe. If we do an MRI scan and look qualitatively at the individual slices, for that kind of patient it would typically be normal, given the normal variance of the structure of that part of the brain." In other words, a physician could probably not see anything in the MRI that was clearly abnormal, because there is so much normal variation in that part of the brain. If, however, it were possible to use a computer to compare the patient's MRI with a probabilistic distribution calculated from 7,000 subjects, some parts of the brain might well be seen to lie outside the normal range for brains. "And if you could compare her brain with those of a well-matched population of other 19-year-old left-handed Asian women who smoke cigarettes, had 2 years of college, and had not read *Gone With the Wind*, you might find that there is an extra fold in the gyrus here, the cortex is a half-millimeter thicker, and so on."

In short, because of the data that it is gathering on its subjects and the capability of isolating the brains of subjects with particular characteristics, the probabilistic brain atlas will allow physicians and researchers not only to say what is normal for the entire population, but also what is normal for subgroups with specific traits. And that is something that would not be possible without harnessing the tremendous data-handling capabilities of modern biologic databases.

Summary

The goal of this meeting was to bring together bioinformatics stake holders from government, academe, and industry for a day of presentations and dialogue. Fifteen experts identified and discussed some of the most important issues raised by the current flood of biologic data. Topics explored included the importance of database curation, database integration and interoperability, consistency and standards in terminology, error prevention and correction, data provenance, ontology, the importance of maintaining privacy, data mining, and the need for more computer scientists with specialty training in bioinformatics. Although formal conclusions and recommendations will not come from this particular workshop, many insights may be gleaned about the future of this field, from the context of the discussions and presentations described here.

APPENDIX A

Agenda

Board on Biology

BIOINFORMATICS:
Converting Data to Knowledge

Date: February 16, 2000
Location: National Academy of Sciences, 2100 C Street, NW, Washington, DC — **Auditorium**

8:00 AM	Continental Breakfast in the Great Hall
8:30	Opening Remarks **Gio Wiederhold,** Stanford University
8:40	Opening Presentation: The Signal Transduction Knowledge Environment **Brian Ray,** American Association for the Advancement of Science
9:00	**Session I:** Generating and Integrating Biological Data A. Methods for data collection **Dong-Guk Shin,** University of Connecticut B. Data characteristics **Stephen Koslow,** National Institute of Mental Health C. Data integration **Jim Garrels,** Proteome, Inc. Moderated Discussion **Susan Davidson,** University of Pennsylvania
10:30	Break

10:45 **Session II:** Interoperability of Databases

A. Design features of interoperable databases
Daniel Gardner, Cornell University

B. Information retrieval and complex queries
Peter Karp, SRI International

C. Definition of data elements and database structure
William Gelbart, Harvard University

D. Novel approaches to achieving interoperability
James Bower, California Institute of Technology

Moderated Discussion
Perry Miller, Yale University

12:30 PM Lunch

1:30 **Session III:** Database Integrity

A. Curation and quality control
Michael Cherry, Stanford University

B. Error detection protocols
Chris Overton, University of Pennsylvania

C. Methods for correcting errors
Bill Andersen, Knowledge Bus, Inc.

Moderated Discussion
David Galas, Keck Graduate Institute of Applied Life Science

3:00 Break

3:15 **Session IV:** Converting Data to Knowledge—Analytical Approaches

A. Modeling and simulation
James Bower, California Institute of Technology

B. Data Mining
Douglas Brutlag, Stanford University

C. Visualization of model fit to data
John Mazziotta, University of California, Los Angeles

Moderated Discussion
Ray White, University of Utah

4:45 Summary
Gio Wiederhold, Stanford University

5:00 Reception in the Great Hall

APPENDIX B

Participant Biographies

Bill Andersen is chief technology officer at Knowledge Bus, Inc. The company is working with the European Media Laboratory Scientific Databases and Visualization Group in Heidelberg, Germany, on the creation of an ontology comprising knowledge of biochemical pathways. This ontology will be used to support multiple activities, including database generation, visualization, simulation, and natural-language processing of textual research reports. This work is being done in collaboration with ZMBH, EMBL, and Lion Bioscience AG and has as its initial goal the comprehensive analysis of *Mycoplasma pneumoniae*. Mr. Andersen's work has been primarily in artificial intelligence and databases. His graduate work at the University of Maryland was on parallel algorithms for frame-based inference systems and on management of large knowledge-based systems. Starting in 1995, while working for the US Department of Defense, he began work on the the automatic generation of databases from computational ontologies, leading eventually to the founding of Knowledge Bus, Inc., in 1998 to commercialize the technology. Mr. Andersen has a BA in Russian language and a BS in computer science from the University of Maryland. He is currently working on his PhD in computer science at the University of Maryland.

James Bower is professor of Biology at California Institute of Technology. His laboratory created and continues to support the GENESIS neural simulation system, which is one of the two leading software systems used around the world to construct biologically realistic neural models at levels of scale from subcellular to systems. As part of the GENESIS project, Dr. Bower's research group has been developing software tools to facilitate access of modelers to the data on which their models depend and access of nonmodelers to model-based analysis of their systems. The GENESIS project also involves a significant educational component, which now forms the basis for many courses in computational neuroscience around the globe. Overall, the GENESIS project is intended to provide a new mechanism for scientific communication and collaboration involving both models and data. Dr. Bower's laboratory has also been involved in the development of silicon-based neural probes for large-scale multineuron recording procedures. These data are critical for the evaluation of network models of nervous-system function. Dr. Bower has a BS in zoology from Montana State University and a PhD in neurophysiology from the University of Wisconsin, Madison. He was a postdoctoral fellow at New York University and at the Marine Biological Laboratory in Woods Hole. He has been at California Institute of Technology since 1993.

Douglas Brutlag is director of the Bioinformatics Resource at Stanford University School of Medicine and professor of biochemistry and medicine at Stanford University. Dr. Brutlag's group works in functional genomics, structural genomics, and bioinformatics. They develop methods that can learn conserved structures, functions, features, and motifs from known protein and DNA sequences and use them to predict the function and structures of novel genes and proteins from the genomic efforts. The group uses statistical methods and machine learning to discover first principles of molecular and structural biology from known examples. They are also interested in predicting the interactions between ligands and proteins and between two interacting macromolecules and are actively studying the mechanisms of ligand-protein and protein-protein docking. Their research approach uses a variety of different representations of sequences and structures. Multiple representations of sequences include simple motif consensus sequence patterns, parametric representations, probabilistic techniques, graph theoretic approaches, and computer simulations. Much of the work consists of developing a new representation of a structure or a function of a macromolecule, applying the methods of machine learning to this representation, and then evaluating the accuracy of the method. The group has developed novel representations of sequence correlations that have predicted amino acid side-chain interactions that stabilize protein strands and helices. They have developed novel algorithms for aligning sequences that give insight into the secondary structure of proteins and developed novel methods for discovering both sequence and structural motifs in proteins that help establish semantics of protein structure and function. Dr. Brutlag obtained his PhD in biochemistry from Stanford University in 1972.

Michael Cherry is head of the Genome Databases Group at the Department of Genetics, Stanford University; School of Medicine Project manager and head curator, *Saccharomyces* Genome Database; principal investigator, *Arabidopsis thaliana* Database; computing manager, Stanford DNA Microarray Database; and co-principal investigator, *Arabidopsis* Functional Genomic Consortium. His group at Stanford is involved with bioinformatics and computational genomics. The group provides two resources: the *Saccharomyces* Genome Database and the Stanford Microarray Database. It provided the *Arabidopsis thaliana* Database until September 1999. The genome databases are service projects for the scientific community that collect, maintain, and distribute information. The group also creates software tools and adds value to these Web resources via curation. The group is involved in various analyses of genomes and their gene products. The third major project is on DNA expression microarrays. It is working with Stanford laboratories on yeast, human, mouse, *E. coli*, *C. elegans*, and *Arabidopsis* microarrays. Dr. Cherry's interests are in integrating and facilitating the analysis of the vast amounts of information in genome and microarray databases.

***Susan B. Davidson** is professor of Computer and Information Science and co-director of the Center for Bioinformatics at the University of Pennsylvania, where she has been since 1982. She got her BS in mathematics at Cornell University (1978) and her PhD in electrical engineering and computer science at Princeton University (1982). Jointly with G. Christian Overton, Val Tannen, Peter Buneman, and Limsoon Wong at Penn, she has developed BioKleisli, a system for integrating biomedical databases that is being used within the Tambis project at the University of Manchester and for several projects in SmithKline Beecham pharmaceuticals. Her current research projects include techniques for automating the development, annotation, and refreshing of biomedical-data warehouses and the use of high-speed networks to connect Mouse Brain Atlas image data with genomic data.

***David Eisenberg** is director of the UCLA-Department of Energy Laboratory of Structural Biology and Molecular Medicine and professor of chemistry and biochemistry in the Department of Biological Chemistry, UCLA. Following a thread of discovery from his earlier work on sequence families and assignment of protein sequences to 3D folds, he is now concen-

---

*Planning Group Members

trating on assigning genome sequences to biologic functions. The new methods that he and his co-workers developed go beyond traditional sequence similarity; they depend on correlation of other properties: correlated inheritance of proteins into species, correlated fusion of domains into single protein chains, and correlated mRNA expression patterns. These methods are intended to guide, complement, and interpret experiments. When applied to whole sequenced genomes, these methods show astonishing power for identifying protein functions, protein pathways, and networks of function. His honors include a Rhodes Scholarship, an Alfred P. Sloan Fellowship, a Guggenheim Fellowship, National Academy of Sciences membership, American Academy of Arts and Sciences membership, the Protein Society Stein and Moore Award, the Pierce Award of the Immunotoxin Society, a Repligen Award in Molecular Biology, Biophysical Society fellowship, and the Amgen Award of the Protein Society.

***David Galas** is chief academic officer of the Keck Graduate Institute in Claremont, CA. He was formerly president and chief scientific officer of Seattle-based Chiroscience R & D, Inc., one of the first biotechnology companies to assemble a full gene-to-drug discovery program. Previously, Dr. Galas served as director for health and environmental research at the US Department of Energy, where he headed the Human Genome Project from 1990 to 1993. He also served as professor of molecular biology at the University of Southern California, where he directed the molecular-biology section for 5 years. Dr. Galas earned his PhD in physics from the University of California, Davis-Livermore in 1972.

Daniel Gardner is professor of Physiology and Physiology in Neuroscience at Cornell University. Dr. Gardner has just published the first comprehensive description of a datastructure for neurobiologic databases. In collaboration with cortical neurophysiologists at several institutions, he is also developing an Internet-accessible database called the Cortical Neuron Database. It will contain electrophysiologic and other information describing cortical neurons and their characteristic responses to somatosensory and other stimuli. Dr. Gardner is using a Common Data Model, designed to serve the needs of interoperability between disparate neuroscience data resources throughout the Human Brain Project and beyond. In addition, he heads the development of the Aplysia database project.

James Garrels is cofounder of Proteome, Inc. Dr. Garrels spent 17 years at the Cold Spring Harbor Laboratory, where his group developed the

---

*Planning Group Members

QUEST system for two-dimensional gel electrophoresis and computer analysis. This was a leading-edge facility in a field that is now called proteomics. In 1995, he cofounded Proteome, Inc. with his wife, Dr. Brooks. They have built a growing business around the production of highly annotated proteome databases using genomic and literature sources. They have comprehensively curated proteome databases for yeast and worm (*C. elegans*), with more species on the way.

William Gelbart is professor of Molecular and Cellular Biology at Harvard University. Dr. Gelbart is also a scientific adviser to the Genomes Division of the National Center for Biotechnology Information and an external adviser to the National Human Genome Research Institute (NHGRI) large-scale human and mouse genome-sequencing projects. Since its inception 7 years ago, Dr. Gelbart has been the principal investigator of FlyBase, the NHGRI-funded database of the genome and genetics of the fruit fly *Drosophila*. Among its other duties, the roughly 30-person FlyBase group is responsible for maintaining the annotation of the soon-to-be-released full sequence of the *Drosophila melanogaster* genome. In addition, it maintains a thorough curation of the *Drosophila* literature and through collaborations with other databases is involved in many projects to provide a rich set of links and relationships with information from other biologic systems. Such database interoperability is one of the major issues facing bioinformatics, and FlyBase is heavily involved in exploring this area. Dr. Gelbart obtained his PhD in 1971 from the University of Wisconsin. He did his postdoctoral work at California Institute of Technology and the University of Connecticut.

Peter D. Karp is senior computer scientist and director of the Bioinformatics Group at SRI International. His bioinformatics work has focused on metabolic-pathway bioinformatics and on biologic databases. He is the bioinformatics architect of EcoCyc and of MetaCyc. MetaCyc is a multispecies metabolic-pathway database. EcoCyc is a pathway-genome database for *E. coli* that integrates information about its full metabolic-pathway complement and its genome. Those data are combined with a powerful graphical user interface. EcoCyc is the first database to describe the full metabolic map of a free-living organism. Dr. Karp has also developed novel algorithms for predicting the metabolic map of an organism from its genome. His work on databases has included development of the object-oriented database system that underlies EcoCyc and work in the area of interoperation of heterogeneous biologic databases. He has organized two workshops in this area and has written several publications on database interoperation. Dr. Karp earned his PhD in computer science from Stanford University in 1989. He was a postdoctoral fellow at the NIH National Center for Biotechnology Information.

***Richard M. Karp** is a professor of Computer Science and Engineering at the University of California, Berkeley and a senior research scientist at the International Computer Science Institute in Berkeley. He received his AB (1955), SM (1956), and PhD (1959) from Harvard University. He has worked at IBM Research (1959-1968), Berkeley (1968-1994, 1999-present), and the University of Washington (1995-1999). The unifying theme in his work has been the study of combinatorial algorithms. He has worked on NP completeness, parallel algorithms, probabilistic analysis of algorithms, randomized algorithms, and on-line algorithms. His current research is concerned with strategies for sequencing genomes, the analysis of gene-expression data, and other combinatorial problems arising in molecular biology. He has received the US National Medal of Science, the Harvey Prize (Technion), the Turing Award (Association for Computing Machinery), the Centennial Medal (Harvard University), the Fulkerson Prize (American Mathematical Society and Mathematical Programming Society), the von Neumann Theory Prize (Operations Research Society of America and the Institute for Management Science), the Lanchester Prize (Operations Research Society of America and the Institute for Management Science), the Von Neumann Lectureship (Society for Industrial and Applied Mathematics), and the Distinguished Teaching Award (University of California, Berkeley). He is a member of the National Academy of Sciences, the National Academy of Engineering, and the American Philosophical Society and a fellow of the American Academy of Arts and Sciences. He holds four honorary degrees.

Stephen H. Koslow is director of the Office on Neuroinformatics and associate director of the National Institute of Mental Health (NIMH). From 1990 to 1999, he served as the director of the NIMH Division of Neuroscience Research. Before that he worked in the NIMH Intramural Research Laboratories and in the extramural programs, where he established the first neuroscience research program. Dr. Koslow serves as the chair of a Neuroinformatics Working Group of the OECD Megascience Forum and as a cochair of the EC-US Neuroinformatics Committee. He has received numerous awards in recognition of his accomplishments, serves on the editorial boards of numerous neuroscience journals, and is a consultant to a number of private organizations and businesses. He received his BS from Columbia University and his PhD in pharmacology from the University of Chicago.

---

*Planning Group Members

John Mazziotta is professor of Neurology, Radiological Sciences, and Pharmacology at UCLA; director of the division of brain mapping; and associate director of the Neuropsychiatric Institute. He runs the largest consortium of the Human Brain Project and is constructing a probabilistic database for brain imaging.

***Perry L. Miller** is director of the Center for Medical Informatics and professor of anesthesiology at Yale University School of Medicine. He has been involved in a number of research projects involving clinical informatics, including work on computer-based clinical-decision support, network-based clinical information access, informatics in support of clinical research, and work as part of the Next Generation Internet initiative. He collaborates with several colleagues at Yale doing neuroinformatics research as part of the national Human Brain Project. He has also collaborated for many years with various researchers to build databases and informatics tools in support of genetics and genomics. Dr. Miller received his PhD in computer science from Massachusetts Institute of Technology and his MD from the University of Miami.

G. Christian Overton was the director of the Center for Bioinformatics at the University of Pennsylvania. He held dual appointments as associate professor at University of Pennsylvania in the Departments of Genetics and Computer and Information Sciences. His work focused on annotation of the human genome through computational analyses and database integration. Database integration, which remains one of the more formidable challenges facing bioinformatics, enables access to vertical information within a species (genome, transcriptome, and proteome information) and horizontally across species to identify orthologous relationships. Dr. Overton received his PhD in biophysics from Johns Hopkins University in 1978. He did his postdoctoral work at the Wistar Institute in developmental biology and earned a master's degree in computer and information science at the University of Pennsylvania. After spending 5 years in the computer industry, he returned to academe to participate in the Human Genome Project.

Brian Ray is senior editor of *Science* and editor of the Signal Transduction Knowledge Environment (STKE). Dr. Ray is responsible for the selection and editing of research papers in signal transduction, the cell cycle, and cell biology. His interest and experience in bioinformatics have developed from his role in the design and implementation of *Science's* Signal

*Planning Group Members

Transduction Knowledge Environment. The Knowledge Environment is a resource for scientists that uses the World Wide Web to provide efficient access to multiple kinds of information, including a large database of information on signaling molecules and their interactions. Dr. Ray earned his bachelor's degree from the University of California, Berkeley and his PhD from the University of Virginia. He did postdoctoral research with Tom Sturgill. Dr. Ray is best known for his work in the discovery of mitogen-activated protein (MAP) kinase, now known to be a member of a class of enzymes that participate in regulation of a broad range of cellular processes from cell division to cell death.

Dong-Guk Shin is professor of Computer Science and Engineering at the University of Connecticut. Dr. Shin's research interests include database interoperability, knowledge discovery from databases, and graphical user interface design for databases. For the last few years, Dr. Shin has been leading a number of research projects related to bioinformatics through funding from National Institutes of Health, National Science Foundation and Department of Energy. He has developed a user-friendly graphical ad hoc query interface that enables computational biologists to quickly learn and examine public genome database schemata and produce semantically correct SQL queries graphically. He has also been developing a graphical data-flow editor that computational biologists can use to integrate a series of data analysis and database querying activities into one seamless data flow. Recently, Dr. Shin has been expanding his previous graphical query editor work so that it can allow computational biologists to express queries against GenBank in any manner they wish. Another current project of Dr. Shin is to develop a database including physiology models, cell images, and biochemical and electrophysiologic data to support the Virtual Cell framework. In 1999, Dr. Shin was the recipient of the University of Connecticut's Chancellor's Information Technology Award. Dr. Shin holds MSE (1981) and PhD (1985) degrees in computer science and engineering from the University of Michigan, Ann Arbor. He joined the University of Connecticut faculty in 1986. During the 1993-1994 academic year, he was a visiting faculty member at the Genome Data Base at Johns Hopkins University.

***Ray White** is the Thomas D. Dee II Professor of Human Genetics at the University of Utah, founding director and senior director of science at the Huntsman Cancer Institute, and chair of the Department of Oncological Sciences at the University of Utah. His research is directed toward the

*Planning Group Members

identification and characterization of genes associated with inherited cancer syndromes. In the early 1980s, his work was instrumental in clarifying the genetic mechanism underlying development of retinoblastoma, an inherited cancer of the eye; his concept provided a paradigm for a class of genes that have come to be called tumor suppressors. Honors have included the 1993 Rosenblatt Prize for Excellence from the University of Utah, the Rosenthal Foundation Award from the American Association for Cancer Research, the Charles S. Mott Prize for Cancer Research from the General Motors Foundation, the National Medical Research Award from the National Health Council, the Distinguished Research Award from the University of Utah, the Allan Award for Cancer Research from the American Society of Human Genetics, the Friedrich von Recklinghausen Award from the National Neurofibromatosis Foundation, and the Brandeis University Lewis S. Rosenstiel Award for Distinguished Work in Basic Medical Sciences. Dr. White earned a BS from the University of Oregon and a PhD from the Massachusetts Institute of Technology. He pursued postdoctoral study at Stanford University and was a member of the faculty of the University of Massachusetts School of Medicine at Worcester before going to Utah in 1980.

Gio Wiederhold is professor of Computer Science at Stanford University, with courtesy appointments in the Departments of Medicine and Electrical Engineering. Dr. Weiderhold's current research focus is on gaining precision in integration of information from multiple autonomous sources. Issues addressed in that domain include the resolution of semantic inconsistencies, effective delegation of processing to remote nodes, simulation for augmentation of results, and providing security and privacy in collaborative settings. His early research contributions included development of real-time data-acquisition technology for medical research (1966), time-oriented databases for ambulatory care (1972), the initiation of knowledge-based research (1977), and the concept of mediated architectures for information integration (1990). During a 3-year leave at ARPA/DARPA, Dr. Wiederhold initiated programs in intelligent information integration and participated actively in the establishment of the National Science Foundation Digital Library initiative. Recent research results include an approach to protect the release of private data in settings where broad access must be granted. Dr. Wiederhold was educated in the Netherlands, started programming there in 1957, and came to the United States in 1958. In 1965, he joined Stanford to direct a computing project for professors Feigenbaum and Lederberg. He obtained a PhD in medical information science from the University of California, San Francisco, and joined the Stanford faculty in 1976. He has been elected a fellow of the ACMI, the IEEE, and the ACM.